THIS BOOK BELONGS TO:

THE WONDERFUL WORLD OF PANDAS

Mimi Jones

Dedicated to all the panda lovers out there.

ISBN 978-1-958985-62-5

www.joeysavestheday.com

A Mimi Book

Pandas are bears! Even though their fluffy faces look different from other bears, they're part of the Ursidae family.

Pandas have an extra thumb. They have a special wrist bone that acts like a thumb to help them grip bamboo.

Pandas have a natural black-and-white camouflage. Their unique fur pattern helps them blend into snowy mountains and shadowed forests.

Pandas are native to China. They live mostly in the mountainous regions of Sichuan, Shaanxi, and Gansu.

China

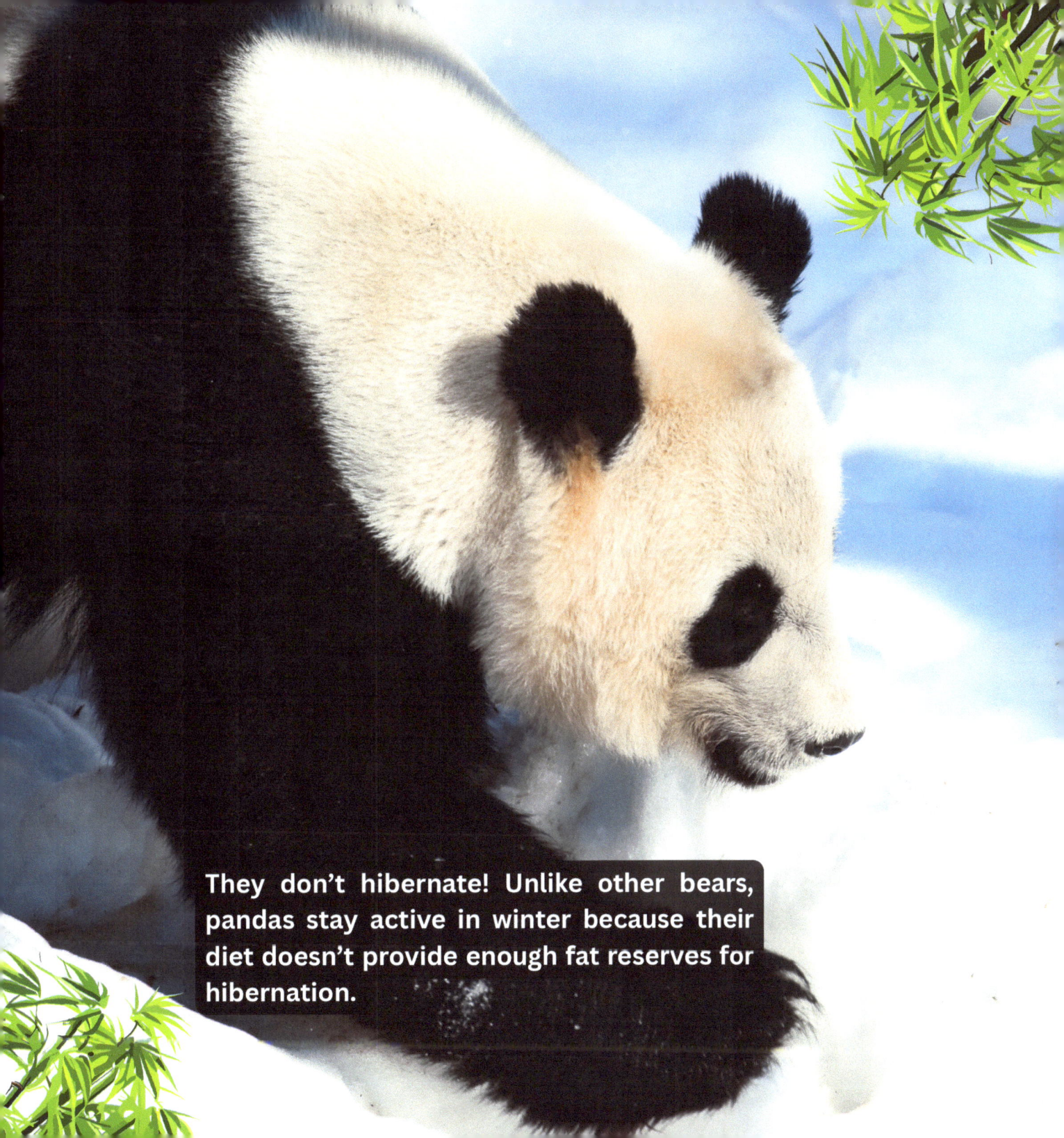

They don't hibernate! Unlike other bears, pandas stay active in winter because their diet doesn't provide enough fat reserves for hibernation.

Pandas are super sniffers! They have a great sense of smell, which helps them find food and detect other pandas.

SUPER!

Bamboo makes up 99% of their diet! They eat as much as 80 pounds of bamboo every day.

99%

Pandas have strong jaw power. Their jaw muscles are so strong, they can crush tough bamboo stems effortlessly.

STRONG

CRUSH!

30

Bamboo buffet. Pandas eat over 30 different species of bamboo.

Pandas are carnivores. Despite their love for bamboo, pandas still have the digestive system of a carnivore and occasionally eat small animals.

Pandas munch all day long. They spend up to 14 hours a day eating!

Munch Munch

14

Pandas drink water, too! They find fresh water in rivers and streams near their bamboo forests.

Pandas grow FAST! In about six months, a panda cub will have its iconic black-and-white fur and weigh around 20 pounds.

FAST

Cubs start climbing early. By three months old, pandas start climbing trees.

Twins are rare in the wild. If a mother panda has twins, she usually cares for only one, but in zoos, humans help raise both.

Panda playtime! Cubs love wrestling with their mothers.

Lazy but lovable. Pandas are mostly solitary and spend their time eating, sleeping, and relaxing.

Mother's milk is essential. Baby pandas drink milk for the first six to eight months of their lives.

6 To 8 Months

Pandas are tree nap champions. They nap in trees or sprawled across rocks, whichever feels comfy.

CHiRP

GROWL

Pandas can make cute noises!
They chirp, growl, and even bleat (like a goat).

BAH!

Water-loving bears! Pandas enjoy playing in rivers and streams to cool off.

Pandas love rolling! Sometimes they roll down hills just for fun.

✨FUN💧

Pandas have been around for millions of years!
Fossil evidence suggests pandas existed at
least 2 million years ago.

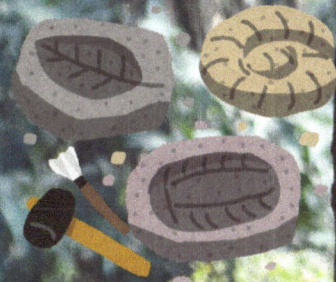

China's national treasure. The giant panda is one of China's most beloved animals.

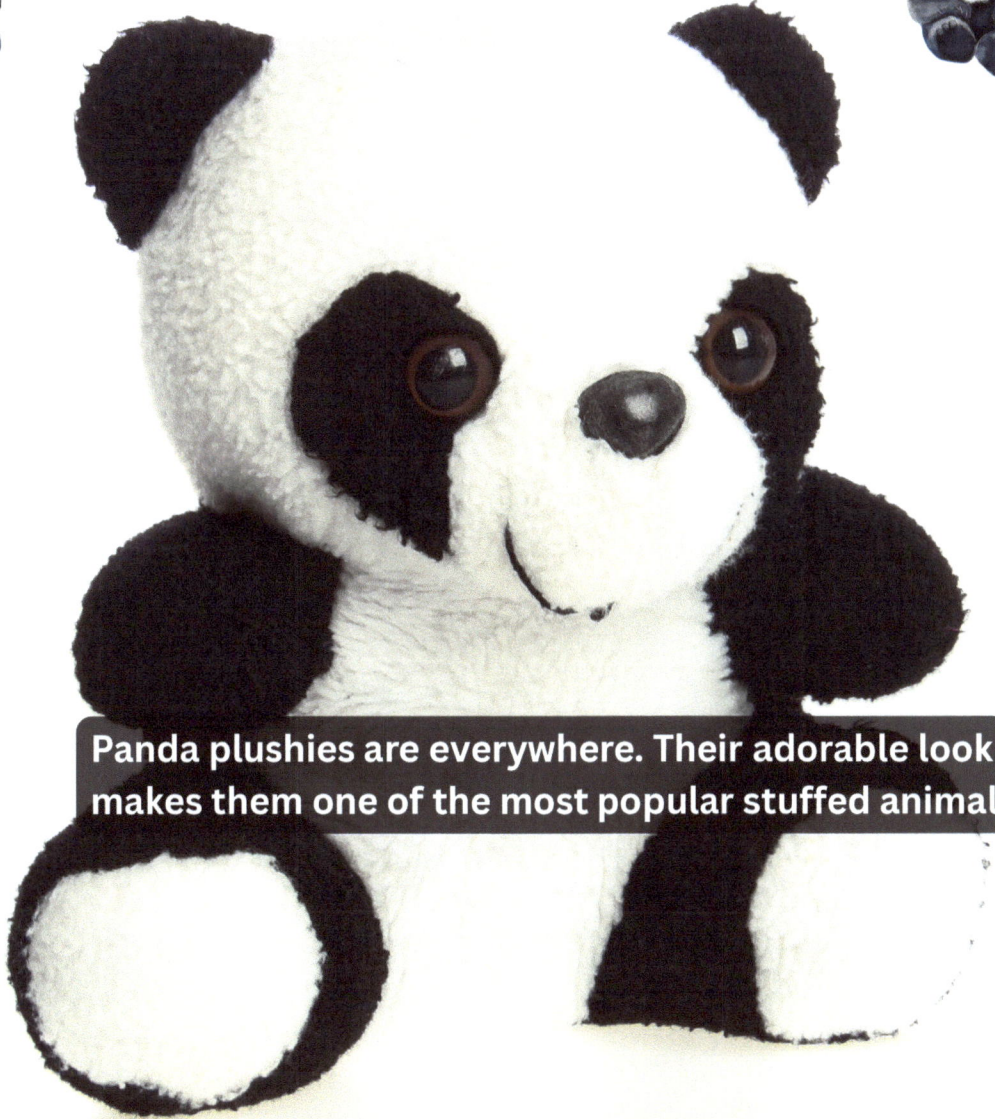

Panda plushies are everywhere. Their adorable look makes them one of the most popular stuffed animals.

Pandas groom themselves like cats. They use their paws to wipe their faces, clean their fur, and keep themselves tidy.

clean up

Their pupils are different from most bears. Instead of round pupils, pandas have vertical, cat-like pupils that help them see better in dim light.

Count the pandas.

Thank you for dedicating your time to read this. I hope you found the information enlightening and that it provided you with some valuable insights.

Bye

www.ingramcontent.com/pod-product-compliance
Lightning Source LLC
Chambersburg PA
CBHW060838270326
41933CB00002B/130